HIGH PERFORMANCE GARDENING

The most fun, productive and organic gardening experience you will ever have!

Lynn Gillespie

Copyright © 2016 Lynn Gillespie

All rights reserved.

ISBN: 1-929709-06-4
ISBN-13: 978-1-929709-06-9

DEDICATION

This book is dedicated to all those gardeners who have struggled to grow a garden. I have felt your pain and together we can garden without the struggle. There is an easier way and I will show you.
This book is also dedicated to my family Tom, Jenny, Mike and Ben who have watched the high performance garden come to life over the last 28 years. They have helped me problem solve through the failures and eaten the successes. This would never have come about without their support and help..

COPYRIGHT

Copyright © 2016 by Lynn Gillespie. All rights are reserved.
No part of this book may be reproduced or transmitted in any form without the written permission of the author, except for the inclusion of brief quotations in a review.

The author and publisher have prepared this book with the greatest of care, and have made every effort to ensure the accuracy, Every effort has been made to make this book as complete and accurate as possible. However, there may be mistakes in typography or content.

Also, this book contains information on gardening and technology only up to the publishing date. Therefore, this book should be used as a guide – not as the ultimate source of gardening. The author and publisher assume no responsibility or liability for errors, inaccuracies or omissions. Before you begin, check with the appropriate authorities to insure compliance with all laws and regulations.

The author and publisher shall have neither liability nor responsibility to any person or entity with respect to any loss or damage caused or alleged to be caused directly or indirectly by this book.

HIGH PERFORMANCE GARDENING

CONTENTS

	Introduction	9
1	Low Performance Gardening History	13
2	Discovering High Performance Gardening	15
3	What is a High Performance Garden?	18
4	Harmony with Nature	25
5	Bug & Disease Resistant	46
6	Weed Free Gardening	49
7	Very Little Time	54
8	Very Little Inputs	58
9	Few Tools Required	69
10	Planning	72
11	Succession Planting	80
12	Extending the Growing Season	83
13	Utilize all the Space	87
14	Create Huge Yields	93
15	Fun & Enjoyable	96
16	Your Life's New Chapter	100

ACKNOWLEDGMENTS

I would like to thank all the people that have lived through the trial and errors as this gardening system developed. I would like to thank Jim Brett for all his support, encouragement and website work. I would like to thank Elaine Brett for editing and encouraging me when things got tough. I would like to thank Kristin Evans for her dedication to this project and all the social media work, editing, and customer service work that she does..

Lynn Gillespie

INTRODUCTION

There are many reasons for growing a garden. Some folks want a garden for the amazing food it creates; others want a garden to save on their grocery bill, still others know that the food quality is in its highest form in their garden. Others want food security; they want to know that they can feed their family should the industrial food system have a hiccup.

(Did you know that there is only 3 days worth of food in the average grocery store? If the trucks were ever to stop rolling the store would be out of food in just a few days.) Still others garden for the sheer joy of creation and connection to the earth.

I would say that most of us grow a garden for more than one reason. I garden because I enjoy seeing the plants grow and I enjoy the amazing foods and health benefits. I don't like to go to the doctor or pay medical bills so I eat my homegrown nutrient dense organic produce. I grow a garden because I love the taste and the satisfaction of creation. Even after 28 years gardening I still come home excited eager to show my family what vegetables I grew. The wonder and amazement is still there after 28 years. I also grow for the thrill of seeing life's creation. I grow a garden for the money savings.

The grocery store is optional for my family. I estimate that I save over $5000 per year just on vegetables. I also like to know that I can feed my family all year from my garden and what I preserve for the winter. It is such a rich feeling to walk into my pantry that is filled with jars of food that I have created for us to eat all year. I plan my garden so that I can cellar, dehydrate, freeze or can the extra food for the rest year.

What are your reasons for wanting a garden?

Good diets always start with nutrient dense vegetables

If you take a look at the major diets and eating styles like vegan, vegetarians, paleo, gluten free, and other diets they all stress eating lots of nutrient dense vegetables and eliminating processed foods. Removing harmful processed foods and correcting our micronutrient deficiencies by eating fresh nutrient dense foods, is very important to so many people. My goal is to teach you how to grow these nutrient dense vegetables in a high performance garden so you can have the very best nutrition and the easiest garden.

Have you noticed that some people have the easiest time growing a garden and other people struggle to grow anything? Gardening seems to be a big mystery to most people and they cannot seem to figure out how to get great results.

The world is changing and many more people are feeling the urge to get back to nature, to grow their own food and reconnect with the earth. The problem is that most people don't know where to begin. Many who are taught to garden are taught by the people who are also struggling and they learn only to struggle in their own garden. This becomes very discouraging and many people abandon gardening altogether.

There is better way of gardening that can eliminate the struggle and help you create the garden of your dreams. We have all upgraded our cars, computers and phones and now it is time to upgrade our gardening system. The new high performance garden systems will take your gardening out of the Stone Age and create a garden that will exceed all of your expectations.

I started in the archaic low performance row garden taught to me by my parents. For years I struggled to grow enough food for my family. I was living the definition of crazy performing one set of actions expecting different results. Once I realized that I would never get the results I wanted from my low performance system I began to search for a different garden system. I tried many methods and created some new methods of my own. After years of trial and error I created a garden system that can produce over 700 pounds of organic high nutrient dense foods in 128 square feet. I was able to produce that amount of food with only a 15 minute per day commitment and I hardy broke a sweat. My goodness this garden is fun!

The good news is that this system is easy to learn and even people with no gardening experience can easily become proficient and get similar results. In fact if you have no gardening experience you will learn this system faster because you will not have to unlearn old low performance gardening techniques.

If you really want to learn to create the beautiful and bountiful garden that you dream about, then you are in luck! As we journey through this book *you will learn all the concepts of high performance gardening*. Your flower and vegetable gardens alike will thrive in a high performance system.

In this book you will discover

- How to create organic nutrient dense food so that you get the very best nutrition.

- How to grow like Mother Nature so that our food will always be organic.

- How to have a small but very productive garden so that you have time to enjoy other activities.

- How to grow disease and bug resistant plants so that you never have to use chemicals.

- How to plan your garden so that you grow lots of food with very little time, expense and tools so that the garden is fun and easy.

- How to have a weed free garden so you don't get backaches and blisters.

- How to have the garden of our dreams without all the work that will be the envy of the neighborhood.

If you are excited to begin your dream garden, then turn the page and let's get started.

CHAPTER ONE
LOW PERFORMANCE GARDENING

It seems that everything has upgraded except for our gardening systems. We don't drive around in old Model T cars, our phones are no longer attached to the wall and our computers are not the size of refrigerators anymore. It is time for us to come out of the archaic low performance gardening systems that were modeled after large scale farms and go with the new high performance systems that are compact, prolific, organic, easy and fun. There is no reason to keep the low performance garden systems anymore let's put those systems in the compost pile.

Low performance gardens are typically grown in long rows, have poor soil, more weeds than you know what to do with, insect and disease problems, low yields and tons of backaches, blisters and heartaches. The most common low performance garden is the traditional row garden.

If you think back a few generations, horses were used to plow. The horse needed at least 30 inches so they could walk between the crops to cultivate. As industrial farming grew, the horse was replaced with the tractor which needed at least 24 inches between rows to cultivate.

When we scaled down to the family home garden, the tractor was replaced by the rototiller which also needs about 24 inches between rows to cultivate. This is the origin of the idea that you need to place your plants at least 24 inches apart in long rows. It is not necessarily what the plant needs, it is what the cultivator needs. So the idea that gardens should be grown in this low performance way came from applying large scale agricultural practices to the family garden. That was the conventional wisdom, mimic the industrial professional farmers. This type of gardening caused our gardens to be way too big and to require too much care.

Our lives are different than the lives of our grandparents. They had the time to grow food in the old low performance row style gardens. For them it was a matter of growing their own food or starving. Life moved at a different pace for them. Today we all have way more to do and trying to squeeze in growing all of our own food on top of it is too much. We don't have time to spend the afternoon weeding the garden. With the new high performance system we set ourselves up for success from the beginning and only spend about 15 minutes per day in the garden and get better results than our grandparents. We can upgrade and skip all the struggle and time it takes to garden. This will make the garden enjoyable not a chore.

CHAPTER TWO
DISCOVERING HIGH PERFORMANCE GARDENING

When I started gardening I began so I could feed my family good healthy organic food. I suffered from food allergies, and so did my young son. I was on a quest to feed my body and my family the best food that I could find. But really my history with gardening goes back farther than that.

My desire to garden began at an early age. I grew up in a suburb of Denver, Colorado – a city girl. My parents were organic gardeners, and I helped with the family garden while I was growing up. My parents had a 1 acre country lot in which they grew traditional row gardens. The gardens consisted of vegetables and prize winning dahlias and iris – my Dad's passion. I didn't know too much about gardening, only that I really enjoyed working with the land, watching the plants grow and if things went well – I enjoyed the produce. My parents also owned a commercial greenhouse for 5 years, growing house plants and bedding plants for commercial accounts.

When I was 19, I moved from the city to the country to marry my sweetheart, who was a farmer who operated a 130 acre farm. My husband, Tom, grows organic hay, grains and silage. With my love of growing I eagerly took on the challenge of growing food for our family.

With my "passed down" – gardening education I had a pretty good idea of how to grow food. Or at least I thought I did! I knew that growing my own food would allow me the control I wanted over what we were ingesting. I was eager and willing to do anything for my family. So for 8 years, I worked and slaved on my HUGE traditional row garden. Maybe you can relate to this struggle. I would spend hours and hours trying to make the soil better, amending, weeding and feeding. I used what was available to me in the garden centers and supplies from the farm. Space wasn't an issue for me and that allowed me to have too much gardening space!

I managed to grow enough food to feed us and put some up for the winter. The food tasted much better than anything I could buy at the grocery store. I knew exactly what was in the soil feeding the plants and ultimately going in to my family's bodies. It was really difficult work and I was relieved when winter came and I could rest. This was the traditional method of gardening that was developed for large farms, but adopted as "the way" by most average backyard gardeners. The pattern was excitement and preparation in the spring, maintenance and hard work setting in by mid-summer, and eventual overgrowth of weeds and burnout by fall. Sound familiar?

After 8 years of this old fashioned row garden, and the predictable pattern of hard work and burnout, I decided that there must be a better way. I began researching and studying every gardening system I could find, and there are so many!

Permaculture, biodynamic gardening, year-round gardening, indoor gardening, raised beds, square foot, lasagna gardening, vertical gardening, container gardening, straw bale gardens, tire gardening, organic gardening, companion planting, watering systems, hydroponics, composting, insect and disease control... Whatever gardening system I had the chance to study I did.

I have studied and tested every system that I thought could work. I've tested countless theories, systems, ideas, and tips. By the end, I began combining the best of each system, separating the facts from the myths and then using only what worked and made the process easier, more successful and more productive. Then I condensed it all down into one new comprehensive gardening system. A system that shortens my gardening time, increases my harvest, allows me to grow organically and protects my family and the environment. Most importantly this new system produced food that is fantastic to eat, all with less space, cost, and time.

Over the next 20 years I made my gardens more productive and much easier. I began a serious organic vegetable operation and I continued to develop my system. My successful vegetable operation, The Living Farm, has 20,000 square feet of organic vegetable production. 9,000 square feet of that are greenhouses where we grow year round in the mountains of Colorado. Little did I know that I was creating the High Performance system that you are learning about today. I was simply looking for a way to make my farm better and more efficient.

CHAPTER THREE
WHAT IS A HIGH PERFORMANCE GARDEN?

A high performance garden is one of the most fun, productive, and organic gardening experiences you will ever have. It is the garden we envision as we look at the pretty pictures in the seeds catalogs. You know what I mean; you have had those dreams too. So why does the garden not turn out the way that we hope? What goes wrong?

Some people have the easiest time growing a garden and others struggle to grow anything. So why is this? Here is the conclusion that I have realized. There are high performance gardens and low performance gardens.

Here is the typical scenario of a low performance garden. You go out in the spring and rototill an area of the backyard or till the old garden spot. Then you add some fertilizer to your freshly turned soil. You are very excited and bought a lot of seeds and plants, because this year you have decided is going to be incredibly productive. You put in your seeds extra thick to ensure a great harvest and planted your plants, and then you watered. Some of the seeds come up and some of the plants that you planted lived. You knew that some of the plants would die and the seeds would not germinate so you put in extra, lots extra. Your garden is large to accommodate all of your plants. In about 2 weeks the weeds begin coming up and you go out and hoe. The next week there are even more weeds. You are bound and determined to win this year so you spend all day in the garden hoeing the weeds. Your back hurts and you have blisters on your hands. But you have won so you feel a sense of satisfaction.

The next week comes and you go way for the weekend and the weeds come up again. The next weekend you have another activity and you miss hoeing. Now the weeds are starting to take over the vegetables. You like the tomatoes and carrots the most so you weed them a little every night but you can't keep up. So, you abandon half of the garden and decide to keep part of it going. The next thing you notice is that the bugs have moved in. The part that you have been weeding is covered in bugs and the plants are looking sick. After picking up some bug spray from the store you douse your plants hoping to save them from the pests. Now you have a toxic film covering the vegetables that you want to eat. Things are going just as bad this year as last year. You finally throw up your hands and walk away, letting the weeds and bugs take over the garden for the rest of the season. I hear versions of this story every year from frustrated gardeners all the stories are very similar.

Correct me if I am wrong, but this is the way that most people are approaching their gardens. The results they are getting and dismal and back breaking. It is no wonder that people quit their gardens every year. How about a different scenario?

Let's take a look at a high performance garden scenario. Every high performance garden starts at the kitchen table. The first thing you do is ask yourself what you want to grow. Then you figure out how many of each plant you want. Because you are in a high performance garden you will know what the approximate yield will be for each plant therefore how many to plant to feed your family. You look up your planting date and know exactly when to start. You can even start up to 4 weeks earlier because you know how to protect your plants from the cold and you do not have to rototill. You have a special prepared soil that your plants love and this soil never needs to be tilled. Your paths go around your grow beds and they are mulched so you can be in your garden after it rains and not get muddy. You feed the soil, plant the garden and cover it with some mulch. All of the plants thrive in your special soil. You didn't over plant because you had a good plan. The special soil and the mulch take care of the majority of the weeds so you spend about 5 minutes weeding the garden. You have saved money by not owning a rototiller or a hoe. The first few weeks go by and you water and then weed for 5 minutes if you can find enough weeds. The next week you go away for the weekend and come home to a garden that is still perfectly weeded. The plants are starting to really grow. The next weekend you have other activities and are way from the garden.

Monday evening you weed for 5 minutes and the garden looks great. You notice that the plants are growing faster and bigger in the special soil and there are still very few weeds. The mulch is getting thin so you reapply another layer.

Now the plants are producing food and you spend 10 minutes twice a week harvesting the food. You also spend 15 minutes on the weekend putting in some trellises because the plants are getting so big. The garden looks amazing without much effort. The most effort you put into your garden is harvesting. The next week you notice that the harvest is getting really big and you are having trouble keeping up with that much food so you start to give food away to your neighbors. The neighbors come over to your house and look at your garden and are amazed at what you have created. The garden produces food until it freezes down in the fall. You clean out the old vines, feed the soil, cover it with mulch and you are ready to start again in the spring. You scratch your head and say "So that is what a high performance garden can do. Wow that was fun! I want to do that again but next year I am going to make it bigger!"

This is the typical story that I hear from people that I have taught to grow in a high performance garden. They always want to make the garden bigger and they are so excited to grow again next year. Is your garden giving you these feelings? If not this book can help change that and help you to begin to make gardening what it is supposed to be easy, fun and productive.

So, how do you change from a low performance garden to a high performance garden?

The way to change from a low performance garden to a high performance garden is by getting an education and becoming proficient in a few new skills. The high performance gardener has a different set of skills than the low performance gardener. The high performance gardener has a much easier time and way more fun in the garden because of their new skills. They will spend way less time in the garden and get way more food because they know how to get the best performance out of their plants.

What is a High Performance Garden?

There are 12 distinguishing characteristics of a high performance garden.

1. **A High Performance Garden works in harmony with nature**. No chemical fertilizers or sprays necessary. The

minerals and organic compost you put into your garden are there to support a healthy soil web.

2. **High Performance Garden plants are disease and insect resistant.** High Performance Garden plants are disease and insect resistant. Because these plants are living up to their full genetic potential they have less disease, are not very appealing to insects and have a higher resistance to frost.

3. **High Performance Gardens are virtually weed free.** The practice of mulching, raised beds and good plant spacing makes weed maintenance less than 10 minutes a week. Say goodbye to weeds, they are a thing of the past.

4. **High Performance Gardens require very little time.** The most time spent in a high performance garden is planting and harvesting. So just 15 minutes a day should cover all gardening chores and feed 2-3 people all season long.

5. **High Performance Gardens require very little input.** No more endless non organic soil supplements. A high performance garden will have the symbiotic microbial relationship that will feed the plants.

6. **High Performance Gardens require very few tools.** You never walk on the soil that the plants grow in and the soil remains soft and workable. No rototiller required. The only tools you may need are a shovel, trowel and a garden hose.

7. **High Performance Gardens are well planned.** The plan for the garden is established before putting any seeds in the ground. In a high performance garden you know when to plant, where to plant and how much to plant to insure the highest performance in your garden.

8. **High Performance Gardens are grown in succession plantings.** Garden beds need to be filled with plants all the time. Succession planting is when one crop is finished another crop is planted in its place. In a high performance garden each succession is planned before you stick your hands in the dirt.

9. **High Performance Gardens extend the growing season.** Knowledge of your localized growing zone, good planning and covering your crops extends your growing season up to 8 weeks.

10. **High Performance Gardens utilize all the space available.** Tight spacing and vertical growing save space increase your harvest and reduce weeds. This characteristic makes high performance gardening truly stand out.

11. **High Performance Gardens create huge yields.** A high performance garden will yield $15-20 per square foot over the course of one season if done right. In 128 square feet (4x32 ft.) with just 15 minutes per day you can grow enough food for 2-3 people and have extra to give away or preserve for the winter.

12. **High Performance gardens are fun, enjoyable and are the envy of the neighborhood.** A high performance garden is beautiful, fun, productive and so enjoyable.

Imagine what gardening could become if your garden met all these criteria! Organic, nutrient dense, weed free, little time, little expense, few tools, well planned, space efficient, longer season, huge yields, incredibly beautiful and enjoyable. Wow, what a great garden! How awesome would that feel?

The high performance garden can provide enough food for you and your family, if you so choose. I love that I am not dependent on the industrial food chain for my survival. I have the skills to be able to grow all my own food if it should become necessary. Gardening is an essential life skill that our ancestors had and our children need. We want the next generation to have these high performance skills so they can enjoy growing food and have time for other activities.

Let's take a closer look at each of the 12 characteristics of a high performance garden..

CHAPTER FOUR
HARMONY WITH NATURE

We are part of nature and the minute that we remove ourselves from this we start to feel disconnected. It is important to our mental and physical health to be connected with nature. Food is the fundamental connection to nature that we all need. It is vital to the health of our cells that we are eating organic nutrient dense foods. The best place to get the most nutrient dense food is from our own garden.

We know that sickness and disease can come from deficiencies in our food. Food that lacks all of the nutrition we need to thrive can greatly affect our health. Without food we die. And without good nutrient dense food I believe that we also die... It is just a slower process with more serious health complications!

Many people have asked the question, "Isn't all produce good for us?"

The answer is no. Not all produce have equal nutritional values. It is hard to tell if the produce that we buy in the grocery store is of good quality and full of nutrients. In essence, we trust our industrial growers to supply our bodies with the nutrient building blocks we depend on for our wellness. And do you think that these growers are looking out for our health or are they growing to produce quantity over quality?

Our food is not the quality that it once was. The reasons for the declining quality are complex. The main reason is that farming has changed. Agriculture researchers have known since the 1940s that yield increases produced by fertilization, irrigation and other industrial farming practices tend to decrease the concentrations of minerals in the plants. This is called the "dilution effect." These techniques give growers higher yields, and consumers get less expensive food. But there is a hidden long-term cost: that cost is food quality and our health.

Not only are minerals lacking, there is a decrease in vitamins as well. Most of the vitamins are only created in the fruit and veggies as they reach full ripeness. So when produce is harvested prior to being fully ripe to arrive at the grocery store at the peak ripeness, it will not have as many vitamins. Now that more food is being transported from further and further away, our problem has increased

The entire decline in food quality happens without the consumer knowing. The end result is the food that is available for us to buy in our grocery stores, organic or not, has lost its nutrition and is affecting our heath.

However, the good news is that getting good nutrient dense food is obtainable by growing your own food – and that is what this high performance gardening is all about!

So, how do we tell if we have more nutritious food? Can this be tested?

The answer is absolutely, YES!

A superior quality fruit or vegetable will have more carbohydrates, proteins, and a wider variety of minerals in the food. When a fruit or vegetable has higher nutrients the density of the vegetable is greater, because minerals weigh more than water. A low quality fruit or vegetable will have fewer minerals and can be watery or mushy.

There is a meter called a refractometer that can measure the amounts of dissolved solids in fruits or vegetables to tell you which fruits or vegetables are more nutrient dense than others. The higher the reading on the refractometer the better quality the fruit or vegetable and the more minerals are available in a plant soluble form for your body to absorb and the better the produce taste. The BRIX index is the scale of measurement that the refractometer uses. The BRIX index ranges from 0-32. The higher the Brix index the more carbohydrates, proteins, and minerals are in the food.

Using a refractometer and understanding a Brix scale will help you in evaluating the quality of your produce. Some people take their refractometer to the produce market or farmers market to decide what to buy and who to buy from. Wine makers have known for many generations that the best wine always comes from high Brix index grapes.

To give you a few examples a poor tomato would have a Brix index of 4, an average tomato would have a Brix index of 6, a good tomato would have a Brix index of 8 and an excellent tomato that would make you jump for joy would have a Brix index of 12.

You can refer to this table to find out the score for different plants on the BRIX scale

	Poor	Average	Good	Excellent
Alfalfa	4	8	16	22
Apples	6	10	14	18
Asparagus	4	6	8	10
Avocados	4	6	8	10
Bananas	8	10	12	14
Beets	6	8	10	12
Bell Peppers	4	6	8	12
Blackberry	4	6	8	12
Blueberries	7	9	12	15
Broccoli	6	8	10	12
Cabbage	6	8	10	12
Cantaloupe	8	12	14	16
Carrots	4	6	12	18

Casaba	8	10	12	14
Cauliflower	4	6	8	10
Celery	4	6	10	12
Cherries	6	8	14	16
Coconut	8	10	12	14
Corn (Young)	6	10	18	24
Cow Peas	4	6	10	12
Cucumber	4	6	8	12
English Peas	8	10	12	14
Endive	4	6	8	10
Escarole	4	6	8	10
Grains	6	10	14	18
Grapes	8	12	16	20
Grapefruit	6	10	14	18
Green Beans	4	6	8	10

Honeydew	8	10	12	14
Herbs (most)	4	6	8	12+
Hot Peppers	4	6	8	10
Kohlrabi	6	8	10	12
Kumquat	4	6	8	10
Lemons	4	6	8	12
Lettuce	4	6	8	10
Limes	4	6	10	12
Mangos	4	6	10	14
Nectarines	6	12	16	20
Onions	4	6	8	10
Oranges	6	10	16	20
Papayas	6	10	18	22
Parsley	4	6	8	10
Peaches	6	10	14	18

Pears	6	10	12	14
Peas	4	6	10	12
Pineapple	12	14	20	22
Plums	8	12	16	20
Potatoes, Irish	3	5	7	8
Potatoes, Sweet	6	8	10	14
Raisins	60	70	75	80
Raspberries	6	8	12	14
Rutabagas	4	6	10	12
Sorghum	6	10	22	30
Squash	4	8	12	14
Strawberries	6	10	14	16
Sweet Corn	6	10	18	24
Tomatoes	4	6	8	12

Turnips	4	6	8	10
Watermelon	8	12	14	16

Let me tell you the story of my onions.

I was chopping onions one day and the tears were streaming. On another occasion the farm students were in the kitchen cutting onions and we were all tearing up even those of us that were across the room. This was beyond the normal reaction to store bought onions. I was curious why we were getting such a strong reaction to the onions. So, I got out the refractometer and tested the onions. The onions read 12 on the BRIX scale. We were 2 points above excellent for nutrient density. We continue to grow high nutrient dense onions and from now on we cut them outdoors to save our eyes!

I use my refractometer to occasionally test the vegetables that I grow here at the farm. They typically range from excellent and beyond on the BRIX scale. The most important test that I do every day is to taste my produce. I will snack on the food that we are harvesting to see if I have enough minerals in the soil. If a vegetable is not over the top in taste, I will re-mineralize the garden bed.

You have a built in refractometer, your taste buds. The higher on the BRIX scale the food is the better the taste. When you consume fruits and vegetables that are nutrient dense you feel like you have just bitten into the best fruit or vegetable on the planet. Your body jumps for joy as you take another bite. This is our experience every day on the farm we eat so well around here!

My family is so used to eating nutrient dense foods that when I serve store bought organic vegetables they can tell. I tried to sneak in some broccoli from the organic market for dinner one night.

After his first bite my son looked at me and said, "This is not your broccoli". I had to admit that I bought the broccoli. The next year I grew more broccoli to put in the freezer so we could have our delicious nutrient dense broccoli all year round. I am glad that my kids can tell the difference between a high nutrient dense vegetable and a low nutrient dense vegetable. The best part is that it is not hard to provide good food for them.

I am amazed at how sensitive the kids are to the taste of the food. I remember when the kids were little we went out to a buffet restaurant for lunch. My son saw some canned peaches in the display and got all excited and put them on his plate. The look on his face when he bit into them was priceless. He looked at me in horror and asked me what they were. I laughed and told him that they are grown in poor soil, picked green, canned in syrup and labeled peaches. He never made that mistake again.

Fruits and vegetable that have a low Brix reading will taste like cardboard, sour or like nothing and can be very mushy. You know what I am taking about. Go to the grocery store and buy a peach or tomato and taste it without any added sugar or salt. It's not the high nutrient taste of a fresh fully ripened tomato or peach. Fruits and vegetables that are low on the Brix index do not give our bodies the nutrients that we need to be healthy. If we consume low Brix fruits and vegetables our bodies will want more food in order to get the right amount of minerals that we need. Thus we tend to over eat. It is common for food manufactures to add sugar, fat and salt to processed foods to trick our taste buds into thinking we are getting something tasty and nutritious

There are 3 factors that influence a high Brix fruit or vegetable.

1. The fruit or vegetable must be grown in the right soil for the plants to get all the nutrients they need.

2. The fruit or vegetable must be harvested at the peak of ripeness. If a fruit or vegetable is harvested before it

matures it will not have as many vitamins as a fruit or vegetable that ripens on the plant.

3. The fruit or vegetable must be consumed or frozen as quickly as possible. Fruits and vegetables will consume their own nutrients once they are removed from the plant. The sooner you eat the fruit or vegetable, the more nutrition it has. Freezing the fruit or vegetable will preserve almost all of the nutrients.

Another great benefit to the plant and to gardeners is that a plant with a high Brix reading is insect and disease resistant. Plants with lower Brix reading are more prone to attack by insects and are more susceptible to disease. Insect infestation is rare in plants with higher Brix values. Higher Brix value plants also appear to be more resistant to mold, mildew, and other plant diseases. The high Brix value plants can also take a frost better than low Brix value plants. Natural selection culls the nutrient poor plants with bugs and diseases, and the nutrient dense plants are left for reproduction. So growing with the intention of producing high Brix plants produces nutrient dense high quality food.

Is it possible to grow the best nutrient dense food?

The answer is yes – by gardening for maximum nutrition and aiming for high Brix value produce. Part of the solution is found in your soil. The soil that a plant is grown in has an immense effect on the plants health and nutrient density. A healthy high functioning soil will produce crops that are flavorful, aromatic and nutritious.

It is easy to grow fruits and vegetables with high mineral and nutrient density using a high performance system. Just imagine the best quality food for only pennies a day.

By providing the right growing environment, your plants will reach their full genetic potential, and provide you with the freshest and best nutrient dense food you can get. The difficult part is to understand what conditions are required to allow your plants to reach their full potential, with optimum nutrition.

So what is the right soil?

Soil quality is one of the biggest barriers to higher crop yields and is the basis of most garden problems and difficulties. I believe that soil is the most important component to a high performance gardening system.

Most people just go out and dig up the back yard and then plant the garden without any regard for the type of soil that they are planting in. This is the #1 mistake of most gardeners. First off, our back yards usually have the worst soil. This is mostly because of the construction of the house. The subsoil from the excavation is now on above the top soil and it is not suitable for gardening. If you want a high performance garden you need soil that the plants can perform the best in. The preferred soil for growing vegetable is sandy loam. The other problem with the soil in our back yards is that it contains weed seeds that when disturbed they tend to germinate.

There is one other factor of the garden soil that makes all the difference in creating nutrient dense vegetables. The ace in the hole is soil biology.

Why is soil biology important to gardeners?

First and foremost, when the biology in the soil is doing its job, your work as a gardener gets a whole lot easier. This is my secret, I garden smarter not harder.

A garden with good soil will:

- Stimulate healthy root development which allows the plant to transport more nutrients to the vegetables. 26

- Will feed your plants automatically so that they get what they need when they need it.

- Will support good microbes and bugs, which will keep the pests that harm your plants under control and will also take care of most common plant diseases.

- Good soil will hold more water and more air so that the plant roots can be healthier.

- Good soil can help a plant survive drought conditions.

- Good soil will house the microbes that will break down minerals found in natural soil that are not found in the chemical fertilizers, which enhance the flavors of our foods and create nutrient dense foods.

- Will reduce or virtually eliminate the need for gardening tools! Especially the rototiller.

The gardener's life is truly a lot easier when your soil biology is balanced. If your soil is healthy your plants will take in the available nutrients. Your plants will be less stressed and this creates a natural prevention and protection from invading bugs and diseases.

The soil food web is made up of an incredible diversity of organisms. They range in size from the tiniest one-celled bacteria, algae, fungi, and protozoa, to the more complex nematodes and micro-arthropods, to the visible earthworms, insects, small vertebrates, and plants. As these organisms eat, grow, and move through the soil, they make it possible to have clean air, water, healthy plants, and moderated water flow

Let's take a close up look at what happens in the soil food web.

Plants take in sunlight and create carbohydrates and proteins. Some of which are then excreted through their roots in the rhizosphere (the area right around the roots).

Bacteria and Fungi are attracted to the roots to consume the excrement and sloughed off cells from the roots.

The bacteria and fungi are eaten by protozoa and nematodes. The nutrients that the protozoa and nematodes don't need are released as nutrients for the plants in the rhizosphere.

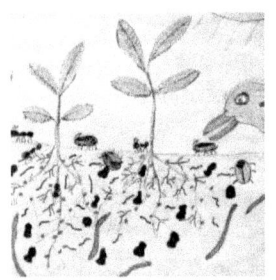

Protozoa and nematodes are eaten by arthropods. Arthropods are the insects that have segmented bodies such as spiders, moths, flies, grasshoppers, butterflies, bees, ants, beetles and more.

Arthropods are eaten by snakes, toads, birds and other animals.

Earth worms that live in the soil eat bacteria, fungi, protozoa and nematodes. If you have a healthy worm population then you have a healthy soil food web.

All dead plants, animals and bugs are eaten by the bacteria and fungi to add nutrients to the soil.

Animal dung is consumed by the bacteria and fungi and also adds nutrients back to the soil.

All of the eating and decomposing has byproducts of nutrients that feed the plants. The soil food web is a delicate structure. If one of the players is removed from the game then the food web can be disrupted and not work.

So here is how all this comes together. The minerals in the soil are broken down by the microbes and are made available to the plants. The plants take up the minerals which then become part of the plants. We eat the plants and the minerals in plant form are readily absorbed by our bodies to become a part of us. This is a healthy food chain that supports healthy humans. When we eat food that is grown from a chemical fertilizer this heathy food chain is broken. The plants are lacking in minerals and so are we.

So how exactly do the chemical fertilizers affect the soil food web?

Chemical fertilizers are made of salts. These are the boxes of blue stuff and the funny smelling stuff on the shelf at the big box stores. They have the NPK (nitrogen, phosphorus and potassium levels listed on the package). These salts draw the moisture out of the microbes and cause them to burst and die.

In addition, chemical fertilizers kill a large percentage of a soil's naturally-occurring microorganisms because chemical fertilizers contain acids, including sulfuric and hydrochloric acids. This acidity adversely affects the soil ph. Thereby changing the kinds of microorganisms and bacteria that can live in the soil. These healthy bacteria would normally break down organic matter into plant nutrients, and help convert nitrogen from the air into a plant-usable form. Plants can survive on the chemical elixir derived from salts but the natural food chain is broken and the microbes no longer receive or give to the plants.

Once the soil food web is destroyed the plants will be dependent on the chemical fertilizer for their survival. Plants can absorb the salt fertilizers and live off of it but this would be like us living off of an IV. Usually the plant will only get 3 of the 17 essential nutrients they need from the chemical fertilizers. 29 When a plant can get "free food" it doesn't form a relationship with the bacteria and fungi. This removes the plant from the soil food web. And without the microbes in the soil, the food chain is broken and the soil structure deteriorates.

It takes almost six weeks for soil to partially recover biologically from a poisoning by synthetic fertilizer. Fertilizer products recommend that you feed the plants every month. This keeps the microbe population under direct attack all the time, without a chance for recovery. With the microbes being killed off, this then causes the soil structure to fail which hinders the soils ability to retain water, air, and nutrients. When a plant is grown under these distressed conditions they are extremely susceptible to damage from diseases, insects and drought.

On the other hand, healthy soil, which is rich in beneficial microbes will, encourage the natural immune systems of plants, help to reduce plant disease, helps the plant to ward off bug attacks and creates the perfect conditions for plant growth. The microbes also help the plant to survive in drought conditions.

Other practices gardeners do to disrupt and destroy the soil food web in their gardens are using pesticides, herbicides and fungicides. Most gardeners also fail to add sufficient organic matter upon which microbes feed and they can starve or move away. Another practice that destroys the microbe populations is compaction of the soil by rototilling or walking on your garden soil.

The good news is that when you have a healthy soil food web you can eliminate your use of chemical fertilizers, herbicides, fungicides and pesticides. When you have a healthy food web system, the good microbes and bacteria influence the plant's health and growth; they also enhance stress tolerance, provide disease resistance, aid nutrient availability and uptake, and promote biodiversity. We find that the incidences of disease and pests are very low in the plants that we grow in healthy soil. We do not use any chemical fertilizer, herbicides, fungicides or pesticides. We do have an occasional break out of bugs but we can eliminate them with some soap or other organic methods.

So why do we not hear about the effects of chemical fertilizers and sprays on our soil food web? These chemicals are advertised to support our gardens – why doesn't anyone talk about how the chemical fertilizers are destroying the soil food web?

Why haven't you heard about the microbes and the soil food web before? Basically it all comes down to money. There is no money in teaching you how to feed your own garden from the materials you have at home. The money is in selling you bags and boxes of potions to grow your plants with. Get your plants addicted to the salt fertilizer so it has no other source of food and you will have to spend your hard earned money over and over again to give your plant its next fix. Then when your plant is weak and sickly you will need the sprays to keep the bugs and diseases at bay, another opportunity to sell you more stuff. Let's stop this cycle of pouring potions and sprays on to your plants and grow Mother Nature's way.

So now what do you do if you have damaged your soil?

Most soil in our back yards is not the best soil for growing plants. Our solution for our gardening system is to create the perfect soil for the plants from scratch and bypass the bad soil that most of us have. By keeping your soil healthy and alive it can support your plants and the vegetables that you consume. This is how professional growers produce the highest quality of plants that you buy in the greenhouses. So why not use their methods?

In our high performance garden system we purposefully maintain the soil to keep the microbes healthy and happy. We use a step by step system to keep our microbes alive a flourishing.

This includes:

1. Never putting anything chemical on the garden or the plants.

2. Feeding the soil to keep the microbes thriving.

3. Turning the soil as little as possible to prevent damage to the microbes.

4. Mulching the garden to give the microbes a cozy home and a source of food.

Using these 4 steps in your garden will change the way you garden forever. This is easy to learn and put into practice and creates an easy organic garden to work in.

Let us take a moment and talk about organic gardening – this is what I call working with Mother Nature to create the right growing environment for your plants to thrive.

So what exactly is organic gardening?

The simple answer is that organic gardeners don't use synthetic fertilizers or pesticides on their plants. But gardening organically is much more than what you don't do. When you garden organically, you think of your plants as part of a whole ecosystem that starts in the soil and includes the water supply, people, wildlife and even insects. An organic gardener strives to work in harmony with natural systems and to continually replenish any resources the garden consumes.

Although the term organic gardening has recently become a trendy buzzword, in fact, it is a concept and practice that has been in use since cultivation of land first began thousands of years ago. It is the way it has always been done. Chemical or conventional gardening is the new comer.

It was only when scientists in the mid-1800s began developing chemical fertilizers and pesticides that the collective mindset shifted toward accelerated growth of crops, more thorough and faster destruction of pests and weeds and an increasing disregard for the environment.

Many people believe that organic gardening takes far more time, money, knowledge and planning than conventional gardening or that it yields measly, tasteless crops that are full of blemishes and bug holes. If you've considered starting an organic garden but have been deterred by these myths, let me share with you my experiences. Your fears may be unsupported

Myth #1: Organic produce is lower in quality.

The quality of the produce depends on the practices of the farmer not whether or not they are organic. A farmer organic or not, can grow plants on depleted soils and get a pretty low quality vegetable. In a high performance garden we concentrate on keeping the soil full of nutrients for the plant to use to make high quality fruits and vegetables. If you grow your own food organically in good quality soil that has all the nutrients that the plant needs you can expect your produce to have: higher levels of essential vitamins, minerals and antioxidants and superior taste. In addition you will have no pesticide residues and no genetically modified organisms. If these distinctions are considered when determining quality, then home-grown organic produce can't be beat.

Myth #2 If you don't use chemical pesticides and fertilizers your yield will be significantly reduced.

This sound like a commercial or ad by a chemical company. In my experience and the same is true for the 60 organic farmers in my valley we get more yields, better taste, the soil improves every year and we have a safer environment for our families. I would never consider using chemical pesticides or fertilizers. Why would I? I can grow better produce without it.

Myth #3 Organic gardens cost more.

Organic fertilizers and bug killers are about the same cost as conventional fertilizers and pest control if you are just comparing the dollar amount. As far as the environment and your health is concerned conventional fertilizers and sprays can cost you dearly in environmental and health damages. If you learn to compost your organic garden can be virtually free.

Myth #4: You have to be an expert or experienced to have a successful organic garden.

This is a complete fallacy. People have been gardening organically for century's way before any chemicals fertilizers or pesticides were introduced. Your very own grandparents probably grew organic gardens. I have lots of students who are complete beginners that have learned my system of growing and they do great in their first season. I also have friends that I have taught and they have transformed their back yards into abundant vegetable gardens.

Gardening is always a journey of learning, and there are certain basic principles that apply. You can just as easily learn an organic method versus a method that uses chemicals. Once you are accustomed to the organic methods you will not need any chemicals to create amazing food.

We grow all our food organically and it is not hard, it is normal. I feel good about having my kids in and around the soil knowing that it is safe for them. It all comes down to having the knowledge of organic gardening instead of the conventional low performance gardening. In an organic garden not only will your garden produce better quality foods it will also keep you in alignment with nature.

CHAPTER FIVE
BUG & DISEASE RESISTANT

Lots of low performance gardens are plagued by insects that destroy their crops. So much money is spent every year on pesticides, and fungicides to keep the plants from being destroyed. This is similar to our National medical plight? We eat really bad food, live in a bad growing environment and when both our systems and the plants show signs of stress in the form of disease our next "solution" is to treat the issue with drugs. I believe for us and for our plants that good nutrition, optimal growing environment and less stress will create a healthy immune system that will take care of most of the attackers.

When growing in a high performance garden we take care of the plant so it has the most ideal environment for it to live in. We make sure that it has access to the 17 essential nutrients that it need for life. Which are the same 17 essential nutrients that we need for life. When the plant has all the building blocks it needs then it can keep its cells healthy and resist attackers.

The 17 Essential Nutrients	
Macronutrients	Micronutrients
Carbon	Boron
Hydrogen	Chlorine
Oxygen	Copper
Nitrogen	Iron
Phosphorus	Manganese
Potassium	Zinc
Calcium	Molybdenum
Magnesium	Nickel
Sulfur	

There are other nutrients in most plants but these are the 17 all plants require for survival.

Other ways to prevent diseases, rot and mildew is to make sure that the plant goes to bed at night with dry leaves. In a high performance garden we protect the plants from too much sun, wind, cold and heat. We keep the weeds down and keep the garden clean. When we put the plants in their more ideal environment then they can thrive.

Another thing that I have noticed in our organic environment is that we have the full soil food web acting on our behalf. Our greenhouses and gardens are full of predators that will eat the bad bugs. We have Lady bugs, praying mantis, snakes, toads that constantly patrol our gardens. The garter snakes give me a fright when I see them but I do respect them and appreciate their contribution to keeping the area free of pests.

After growing for 28 years I have learned how to deal with the few pests and diseases that can harm the plants completely organically. You can learn how to organically eliminate pests and diseases just like you can learn how to do the same with chemicals. The route you take will depend on your education. For the most part when the environmental and nutritional needs of the plant are taken care of the incidences of disease and pests are low. To take care of the little problems that do occur all that is needed is some knowledge about organic and natural ways to keep the pests and diseases at bay. You can begin your organic pest and disease control on by reading the many blogs detailing our methods at www.thelivingfarm.org

Sometimes you will get a pest invasion because you live next to a garden or orchard that is not being cared for and their plants are stressed. Or a pest is just moving through the area and you will pick up some strays from the migration. An observant gardener will keep a watchful eye out for invaders and eliminate them before they can multiply and take over.

When your garden is small it is easy to keep an eye on everything. In a high performance garden the garden is planned out carefully and this helps to keep the garden small and productive. It is easier to maintain and watch for invaders in a small garden. When I had my row garden it was over 3000 sq. ft. The high performance garden that I upgraded to is only 700 sq. ft. and I can grow more food in the small garden than the large garden and it is easier to patrol.

One other thing that I have found out is that a plant that has a high brix rating will be more frost resistant. The plant will have more vital nutrients in its veins and these nutrients act as an "anti-freeze" and protect the plants so they can survive at a lower temperature. My lettuce plants survived a 12 degree Fahrenheit night and I was able to harvest them the next afternoon. Sometimes just a few degrees protection can make the difference between survival and starting over.

When you grow in a high performance garden you do not need a PhD in insects to raise a garden. All you need are some observation skills, high performance garden techniques and a few natural treatments will usually take care of the insects and diseases. A high performance garden is a healthy and vibrant garden that is so easy to maintain and enjoyable to be in.

CHAPTER SIX
WEED FREE GARDENING

Weeds are probably the most annoying and time consuming part of gardening. I think more people quit gardening because of weeds than any other reason. If we want to have a high performance garden then we need to find ways to eliminate weeding. There is no reason to spend hours a day in the hot sun pulling weeds that will come up again days later. Weeds use up the water and the nutrients that the plants need to thrive and they end up robbing you of your time and your harvest.

There are 5 methods I use in my high performance garden to eliminate the need to do hours of weeding.

1. The right soil

2. Raised beds

3. Plant spacing

4. Mulch

5. Vigilance

The Right Soil

At this point you know that the right soil makes all the difference to the plant's growth and to maintaining a thriving microbe population. The right soil will also affect the weeds. In our high performance garden we do not use any native soil from our yard. We manufacture a sandy loam soil that has no weed seeds. You see, native soil all contains a certain amount of weed seeds. These seeds can lay dormant until you disturb the soil then they will germinate with the next watering. When you start fresh with soil that you create you can start with a clean slate. If you start with no weeds and never let any weeds go to seed in your garden or around your garden than you remain virtually weed free all the time.

Raised Beds

I love my raised beds. Can I say this again? I LOVE my raised beds! This is one of the most important parts of the high performance gardens system. In your garden you will have about half the space for the raised bed and half the space for the aisles. The aisles can be crafted in such a way that you will not need to weed them. The aisles can be covered with weed barrier, gravel, wood chips, mulch or it can be your lawn that you just mow or weed whack. Half of the weeds are eliminated by not having to weed the aisles.

In a low performance row garden, the typical practice is to broadcast all the fertilizer all over the garden and rototill it in (which disturbs the soil food web and activates the weed seeds).

Then space is left between the plants for walking in the garden. All this space needs to be weeded and the weeds have just been fed and will be watered regularly along with the plants. This is a great way to get a huge crop of weeds. This is why people abandon their garden by midsummer because they could not keep up with the well fed and watered weeds. By late summer most of low performance gardeners have abandoned their garden. At that point the weeds have gone to seed and the process is ready to start all over again in the spring. This is a great way to set yourself up for failure. High performance gardens break this cycle of growing great big weed crops by only nurturing the plants and not your local horde of weed seeds.

Our high performance gardens are raised beds with gravel or mulched aisles. If we find weeds in the aisles we will use the weed whacker to keep them under control. Or we add some mulch to smother the weeds.

By having aisles that are well mulched we have eliminated half of the weeding. What about the other half? The grow beds are all the weeding you will need to do. Very few weeds will come up in a grow beds with new fresh made soil. We will get a few weed seeds from our mulch, the air and our irrigation water. These weeds can be easily eliminated by hand. In the raised beds as an extra precaution you can put weed barrier under the beds to keep the grass and perennial weeds from coming up into the bed.

Plant Spacing

In the raised beds we plant the plants very close together. This works because we do not need to walk in between the plants in the grow beds, we walk around. When the plants are tightly spaced they cast a shadow on the soil and this lack of light also helps to eliminate weed seeds from germinating.

In a high performance garden we plan the garden before we start. We know how many of each plant that we are going to grow so we don't have to over plant. This reduces the size of the garden to be exactly what is needed for your family and no bigger. With a smaller garden to take care of we don't have to maintain a large plot so the weeding is also reduced.

Mulch Magic

This is my favorite way of keeping the weeds out. We mulch all of our gardens. When the plants are really small we mulch with grass clippings about ¼ inch thick. As the plants get bigger we add more mulch every few weeks until the mulch is over 1 inch thick. For the larger plants we will mulch up to 4 inches or more. You can mulch with different materials. My personal favorite is the grass clippings. When you apply organic grass clippings when they are green, they will interweave and form a tight mat that is hard for the weeds to push up through.

Mulch has several other benefits. Mulch will reduce the amount of watering you need to do by about half if you have your mulch layered thick enough. Mulch not only saves water and reduces weeds it also provides food for the microbes and worms. Mulch makes the garden look awesome and perform wonderfully.

I have mulched with straw but there are several problems with using it. First it is too big to really fit tightly around the plants when the plants are small. If the mulch doesn't fit tightly or is not thick enough you will get weeds coming up through it. It is fine if you are using it in the 4-6 inch range. The other problem with straw is that it still has seeds in it and you can get cereal grains growing in our garden from the straw. I do not like to weed the grains so I don't use much straw as mulch.

Vigilance

The last defense against weeds is to look at your garden every week and pull up the weeds that did make it. There will not be many but it is important to get these weeds so they do not go to seed. For each weed that I pull there are hundreds that I will not have to pull. So when you are weeding you can count by the 100's in potential weed savings!

The great thing about the raised beds is that we only have to weed the bed not the aisles. So the time it takes to look over the garden is less. Also with superior planning that the high performance gardeners do we will have a smaller garden and less time will be required to look for weeds. By the time you apply all of these methods, weeding becomes a thing of the past.

CHAPTER SEVEN
VERY LITTLE TIME

When you garden in a high performance garden it takes very little time to grow amazingly high yield crops. Sometimes I am a little sad that there is not more to do in the garden because I like to be there. When you run out of things to do then take a chair and sit in your garden and simply enjoy it!

We all have busy lives and a garden that requires a lot of time is not going to work in our modern lifestyle. It is important that the garden works for you or you will not garden. Luckily for us, all systems tend to improve over time and gardening is the same. With the advent of the high performance gardens we can organically get a big return with little time and create high yields of the best nutrient dense foods. It is incredible that we can garden this way! I know this is possible I have done this for over 20 years. I can hardly contain myself enough to continue writing because I am so excited to share with you the secrets to high performance gardening. I want to share this with everyone that I meet so that we can share in the joy of an abundant garden together. I love to Imagine what this world would be like if everyone had a super productive easy garden. What do you think life would be like if we all had a high performance garden? Now you can hardly sit in your chair too!!

So let's look at the ways that the high performance garden will require very little time. The components that are saving the time are: planning, raised beds, reduced weeding, reduced size, growing style, enjoyment and fewer trips to the grocery store.

Planning

As a high performance gardener before I plant a single seed I spend the time figuring out what my family and I want to eat. No sense in growing things we are not going to eat or growing too much food. We need to figure out how many people we want to feed and what they like to eat. From there we can determine what we want to grow. Armed with this information we can control the size of the garden thus saving time because the garden is just the right size. Low performance row gardens tend to be huge because the gardener is never sure which crop will succeed so they must over plant and take care of the extra space this requires.

Raised Beds

The raised beds will save us time by reducing our weeds. We covered weeds in the last chapter but we do save tremendous amount of time when we maintain our weed free garden. In a raised bed garden you will find that you will save half the amount of time you would spend in a garden that is not in raised beds. There is half as much to take care of because of the aisle and bed system. The raised beds also require ½ the amount of fertilizer (compost) and ½ the amount of water that an in ground garden uses; which also saves us time and money.

The defined edges of a raised bed garden save you time because you know where to stop. It is much easier to take care of a garden with defined edges. These are perfect gardens for kids because they can see the end of the work.

The raised beds can be tended to in small increments of time. You can do a few minutes in the morning or a few minutes at night or a little bit longer on the weekend. You can add your garden chores to other activities you do during the day. Say when you let the dog out in the morning you can water. Or when you finish the dishes at night you could mulch. You will get your garden chores done without feeling overwhelmed by the garden.

Enjoyment

I know this seems like a strange way to explain saving time and it may not really "save" time. But have you noticed how time flies when you are having fun? A garden that is enjoyable will take less time because you will be having fun!

Planting and harvesting take up most of your time in a high performance garden. My favorite part of gardening is the harvesting. When it is dinner time I will grab a basket and head out to the garden. I will put in the basket whatever is ready to pick and that becomes dinner. This saves me so much time because I don't have to plan much for dinner and I don't have to go to the store to get ingredients. The planning was done when I decided what to grow. It amazes me how much time it takes to go grocery shopping. The stores are so big that it takes forever just to walk around and get your stuff. It is much easier to just pop out to the garden and pick the produce that is at its peak ripeness.

When the kids were younger (and even now) they would play a game during dinner to see how much of the food came from the farm. Sometimes the only ingredients that came from the store were things like salt and oil. This game raised awareness for my family that we really could grow enough food to feed ourselves. It is so much fun to do together and it helps the kids to realize that it is possible to be self-sufficient.

If I have extra vegetables at dinner, I will prepare them for the freezer so we can have those vegetables again in the winter. This is a great way to get some food preserved and also to cuts down on the time and money spent at the grocery store.

The high performance garden is so enjoyable it becomes a wonderful addition to your day. You will look forward to your 15 minutes that you dedicate to spending in your garden. The planting and harvesting in your garden beds will no longer be a burden but a treat! This may become the best 15 minutes of your day. I enjoy this type of satisfaction when I garden I want to help you to get your gardening skills to the same level so you too can have a joyful time in the garden. To give the gift of joy and satisfaction to others is the ultimate gift I can give myself. The fun part is you will experience this same feeling once you have learned these skills and then shared them with others. Your community, gardeners and non-gardeners alike, will notice your amazing garden and they will want to know what you are doing. You can count on being asked how you make your garden grow!

CHAPTER EIGHT
VERY LITTLE INPUT

The marketers of garden fertilizers and sprays want you to get your plants addicted so you will continue buying their products. They are not going to like it when I tell you that once you are composting that the only inputs that you are going to need will be to buy some minerals.

In the raised bed system, you are only nurturing your plants and microbes with the compost and minerals, this reduces the amount of compost and minerals you add by about half of what a low performance garden requires. It is nice to only have to feed the beds and not the weeds in the aisles.

There are as many ways to make compost as there are gardeners. It is easy, cheap and convenient. If you are creating your own compost from the materials you find in your house and yard, this makes your food practically free. Your first few compost endeavors may not go as planned and each yard and mix of materials will be a little different. I challenge you to keep at it until you can compost anything the first time. Make this your personal challenge for the year. Earn the title Super Composter!

With just a little knowledge and a few tools, you can make all the nutrient rich compost that your garden will need. I was watching Joel Salatin on YouTube the other day, and he said that if all the farm land in the world would add 1% organic matter to the soil, we could sequester all the carbon in the atmosphere that has been put there since the Industrial Revolution. Wow, think about this. If we make compost and put the organic matter in the soil, we are helping to reduce the global warming situation. This is even better than changing out our lightbulbs to reduce global warming.

Compost Containers

There are zillions of gadgets on the market for holding compost. For some people it will be easier to buy a ready-made container than to make one. Compost tumblers are the easiest way to make compost fast. The tumblers range in price from $100 to $400. You can start small and work your way up to a bigger tumbler if you need. It is possible to make compost in a pile on the ground but it takes a lot more work, it is usually smelly and animals tend to dig in the ground piles unless you cage the pile. But, it is a free way to make compost.

Easy Composting

If you do a search in the Internet about how to make compost, you are going to get a million different answers. Compost is hard to mess up – if you follow the rules, you can't go wrong. Many different ingredients will work. There are very technical ways to make compost and there are simple ways. We will start with a simple recipe and if you want to perfect compost making, then go online and start reading. Here is the simple way. To make great compost the pile needs 5 basic ingredients.

It needs a source of carbon. This includes our "brown materials", such as dried grass, hay, shredded newspaper, dried leaves, and sawdust. Leaves will add trace minerals (aka "flavor") from deep in the ground. Do not add the leaves of black walnut, magnolia, hemlock, eucalyptus, juniper, pine, oleander. The carbon material is food for the good fungi.

Compost needs a source of nitrogen: this includes kitchen scraps, fresh grass clippings, fresh plant scraps and manures. The nitrogen or "green" ingredients are food for the good bacteria. The manure of cats and dogs can be harmful to humans if composted, so leave them out. Egg shells are great, they add calcium. Coffee grounds are fine too.

Compost needs adequate moisture. Not too wet and not to dry. The way to test if your compost is to wet or dry is to squeeze a hand full and if you can only get a drop of water to come out it is just right. If you can squeeze lots of water out it is too wet. If you get no water and your hand is dry then it needs water.

Compost needs oxygen for the good microbes to breathe.

You will need compost starter, which has microbes in it to inoculate your new pile. You can buy commercially made starter, which might be a good idea for the first batch. After that, you can use some already finished compost to activate your new pile.

Lack of water and oxygen are the main reasons why people fail to make compost. Be sure you have proper moisture and stir the pile twice a week. The smaller you chop the ingredients the faster it will compost. Try to get a balance of carbon, nitrogen and kitchen scraps. Then, add water and stir to get enough air. Put the carbon, nitrogen and kitchen waste in layers. Always cover the kitchen scraps to help keep the critters out.

Do the Ingredients Need to be Organic?

Everything that I put in my compost pile is organic. If you have pesticide or herbicides residues on the scraps that you put in the compost, then you run the risk of harming the microbes and potentially harming the vegetables or yourself. I heard this story where the cows were eating feed that had herbicides in it. When the cows pooped, the poop killed the pasture grass. I do not want to put anything in my compost pile that could be dangerous. I will only use manure from animals that have been fed organically and had no GMO feeds. Also, I will not use any grass clippings that have been treated with chemical fertilizers or herbicides.

One year our town was giving away truckloads of leaves that they vacuum up from the streets. I thought that would be a great way of getting some extra compost material. It wasn't such a great deal. I got leaves but they came with beer cans, glass, and a dead cat. I decided not to put that mix in my compost pile! So, be selective about what you put in your compost.

Composting Process

Here is basically what is happening in your composting process. The carbon and nitrogen materials are being converted into earth through the microbes. The good bacteria and fungi are breaking down the materials. The microbes need an environment that is conducive to their work and growth. The good bacteria and fungi need oxygen and moisture to do the conversion. The bacteria create the heat in the compost process. The heat is good because it can kill weed seeds and bad pathogens. The heat also helps the microbes to increase their population. If your compost is lacking in oxygen, then bad, anaerobic bacteria can grow and create a stink.

A good compost pile needs a carbon to nitrogen ratio of 25:1. You can get into some very complex equations to get this ratio right. We just need to get in to the ball park. For your "brown" stuff use leaves, dry grass or hay, sawdust or shredded black and white newspaper with soy ink.

For your "green" stuff use grass clippings, kitchen scraps (no meat or grease) or fresh plant material. A good ratio to get the carbon to nitrogen in the ball park is 1/2 brown stuff, 1/4 green stuff and 1/4 kitchen scraps, by volume. If you don't have enough kitchen scraps, then use 1/4 more green stuff. Try not to make this too hard on yourself. If you get the recipe close, the microbes will do the work. You are looking for a pile at least 24-36 inches tall and wide.

Put all the ingredients in your tumbler or in a pile. If it is moist enough, it should start to heat up within 24 hours. If it is not doing anything, then add a little water. The outside air temperature can have an effect on the start if it is below 40 degrees. If it is cold, give it a few more days. For your tumbler give it 5 turns every other day. For your pile turn it every other day and rotate the outside stuff to the inside. The compost should be done in 2-3 weeks.

One way to test if your compost is done is by the smell. If it is ammonia smelling, then the bacteria have not finished their job and the compost needs to "cook" some more. Once the bacteria get finished, then the fungi really get going. Once the fungi are done, the compost will smell earthy. Then you know it is done. Another way to tell that it is done is when you can't recognize the ingredients that went in to the pile. This compost is teeming with microbes that are perfect for your garden or lawn. This compost is also what you can use to make compost tea.

If you need to make compost now, get 3 bags of leaves or a bale of straw and a 50-pound bag of alfalfa meal or 3 bags of fresh grass clippings and start a pile. This should make compost in 2-3 weeks. It may take you several tries to get really good compost. If you get a funny pile, just start over. It is important to learn how to make good compost, and it is really easy once you figure it out.

I don't know if you noticed but we are working with the same soil food web in the compost pile as we are in the garden. Mother Nature's way of feeding her plants. This is a beautiful and simple system that can take care of your garden as long as you don't harm the microbes with any chemicals.

An easy and cheap compost container is to get seven shipping pallets and wire them together. I like to use two compost bins at a time. One bin will hold the compost and the second bin will be used when the compost needs turning. You just take down one side from the full compost container and shovel the compost into the empty container. This aerates the pile and helps it to break down faster. No matter what you use for a container be sure that you can stir the pile.

How Much Compost Should I Put On the Garden?

Everyone wants to know how much compost or fertilizer to put on the garden. The fertilizer companies have their fancy ads on the TV telling you to buy this and that to make your garden great.

Let's take a step back in time to see how fertilizing was done in the good old days, before fertilizer companies and TV. You find the richest, most fertile soils in the forests, why is that? Every year the forest floor gets a new layer of leaves and organic matter. The worms and microorganisms come along and digest the leaves, turning them into rich compost for the trees to feed upon. In the garden we should try to mimic nature, by making our own compost we are creating our own rich soil.

In my garden, I add about 1-2 inches of compost to each cinder block unit in the beginning of each growing season. The great thing about compost is you cannot burn the plants with it. If you get too much, the nutrients will just sit and wait until the plants need it. With a good compost heap you can sustain your garden forever. It reminds me of having a sourdough starter.

Here is a reference chart of how much carbon to nitrogen there are in some of the more common compost materials. Just use it as a reference; I would not spend the time to calculate out all the ingredients. Remember, if we get it close, then the microbes will be happy.

Carbon Compost Ingredients	C:N	Nitrogen Higher Ingredients	C:N
Shredded newspaper*	170:1	Fresh grass clippings	15:1
Straw	75:1	Weeds (with no seeds)	30:1
Shredded cardboard	350:1	Kitchen scraps	20:1
Dried leaves	70:1	Coffee grounds	20:1
Old hay	55:1	Manures	10-20:1
Sawdust	400:1	Freshly cut hay	25:1
Wood chips	400:1	Seaweed	20:1
Small branches/twigs	500:1	Alfalfa	12:1
Paper towel	110:1	Hair/fur	10:1

Carbon Compost Ingredients	C:N	Nitrogen Higher Ingredients	C:N
Tissue paper	70:1	Fish emulsions	8:1
Wood ashes	25:1	Blood meal	4:1
Dried grass clippings	50:1		

*with soy ink (no colored paper)

Here is a list of the best compostable items to put in your pile -

- Alfalfa, sawdust (in small amounts)
- Animal fur, seaweed and kelp
- Vegetable and fruit scraps
- Shredded cardboard, shredded newspapers
- Aquarium plants
- Bird cage cleanings, feathers
- Guinea pig, gerbil cage cleanings
- Blood meal, bone meal, hoof and horn meal, fish bones, meal
- Tea bags and grounds, coffee grounds
- Bread crusts, cooked rice, egg shells
- Theater tickets, Kleenex tissues, envelopes, grocery receipts

- Paper napkins and towels, post-it notes
- Brown paper bags, burlap coffee bags
- Tofu, tossed salad
- Cattail reeds, clover, winter rye, straw, hay
- Vacuum cleaner bag contents, lint from clothes dryer
- Wood ashes (in small amounts)
- Cover crops, wood chips, grass clippings, leaves
- Dead bugs, worm castings
- Electric razor trimmings, hair, fingernail clippings
- Expired flowers, houseplant trimmings
- Farm animal manure
- Freezer-burned food (non-meat)
- Limestone, green sand

Here are all of the things to avoid placing in your compost pile.

- Cat or dog feces
- Human feces
- Meat or fat or bones
- Colored newspaper

- Pine needles
- Black walnut leaves
- Wax
- Plastic
- Weed seeds
- Dairy products
- Oak leaves
- Waxed cardboard
- Large pieces of wood
- Pesticide treated plant material
- Grease or oil
- Magnolia leaves
- Hemlock leaves
- Eucalyptus leaves
- Juniper leaves
- Pine leaves
- Oleander leaves

Now is your time to get your compost pile going. Go get some brown stuff, some green stuff and give it a whirl. If it doesn't work out the first time try, try again until you get it. Each batch you will learn something new. Before you know it, you will be a compost superstar! It is well worth the time to learn this skill.

That is all for the inputs in the garden. Because we use a special sandy loam soil and we don't walk on the beds we will not need gas for the rototiller. We don't even need the rototiller! Little time, little supplies, no rototiller or hoe makes the high performance garden really easy.

Speaking of easy, let's take a look at the tools used in a high performance garden system.

CHAPTER NINE
FEW TOOLS REQUIRED

Hey folks this is going to be another short chapter!

The best way to figure out a low performance gardening system is if it starts with "go out and rototill" or "dig down 12 inches". This is your first clue that you are in for rough waters ahead.

Gardens that do not need to be rototilled are my favorites. I am about 5 ft 1 and 100 pounds and the 400 pound rototillers run me instead of me running them. When I was in a low performance system I would bake my husband brownies to get him to till the garden but we wouldn't be able to start until the ground had dried out. This messed up my planting schedule every year. I can grow way more now because I can walk down the aisles of my high performance garden any time even after a rain and plant something. My planting timing is not dependent on the ground to dry out enough to rototill.

So whose bright idea was it to till the soil before we planted our garden?

If you think back to when man started farming a horse was used to pull implements through the soil to plow. The horse needed at least 30 inches between rows to walk thru the crops to cultivate. As farming grew in scale, the horse was replaced by the tractor and it needed at least 24 inches between rows to cultivate. As we scaled down to the family home garden, the tractor was replaced by the rototiller which also needed at least 24 inches between rows to cultivate. This is where the idea of placing your plants at least 24 inches apart in long rows came from. It is not necessarily what the plant needs it is what the cultivator needs. So the illusion that gardens should be grown this way really came from applying large scale agricultural practices to the family garden.

Here is how I prepare high performance garden in the spring. I had cleared my garden last fall by removing old plants and mulching the entire garden with leaves. So in the spring I will take off the leaves and add some compost and minerals. Once that is finished I then lightly turn the surface of the soil to mix in the compost and minerals with my hands. Now my garden bed is ready to be planted. I can prepare the whole garden bed at once or I can just prep the section that I want to plant.

I am glad that we can garden without having to rototill or use other "normal" garden tools. If you own a hoe then you will more than likely have to use it. Yikes! I don't need a hoe or dandelion digger. There are hardly any weeds and the ones I see I get when they are tiny and pull them with my fingers or even easier, smother the weeds with some mulch.

With my special soil mix I can turn the soil with my hand if I want to. This is can be a little hard on the finger nails so I use a trowel. To rake I simply run my hand over the surface and we are done.

Not needing many tools can really save money and you can start a garden without the expense of a rototiller.

I love easy gardens, I plan on gardening until the day I die and with this system I can!

CHAPTER TEN
PLANNING THE GARDEN

To get a high yield out of your garden you need to have a good plan. It is possible with the right growing system to know the approximate yields you will get from each plant and when it will produce. For our 20,000 sq. ft., vegetable operation we have an Excel spread sheet that walks us through every step. When the sheet is filled out we know when to plant what varieties, how many of each plant, how early in the year we can get started and how late in the year we can grow.

Once you have a well-planned system all you need to do is enter the data and you will be ready to go for another season. You can even figure out how many plants you will need if you want to preserve food for the off season. This sounds complicated but it is really simple once you have your system and you do it a time or two. I have heard of several apps you can get for your smart phone to figure all this out but it only takes a few minutes with a piece of paper and a pencil. Do you really want to stare at another computer screen? Let's opt for a more low tech solution for the garden. Once you have the plan drawn you can keep it in your garden journal to look at in years to come.

Here are the key pieces of information that you will want to gather so you can plan your perfect garden.

Your Zone

The United States and many other countries are divided into planting zones; these planting zones approximate how cold and the length of winter in your area. From the zone charts you can determine when your average last frost date will occur and when your first frost will occur in the fall. Once you know when your average first and last frosts will occur you can determine how many frost free days you will have for growing your garden. This information helps you to know when to start planting. Here is a link to an article on zones if you are unfamiliar with your planting zone. http://thelivingfarm.org/project/planting-zone/

High performance gardeners take this information to figure out when to start their cold weather and their warm weather plants. The cold weather plants can go in to the garden 4 weeks before their last frost date and can be in the garden 4 weeks after their average first frost date. This is how we get more production from the same garden.

What to Grow

Once you determine how many days you have to grow each type of plant, then the real planning can begin. The first question you should ask yourself and family is what we want to eat. Make a list of all the things that you want to eat then figure out if they can grow in your area. The zone chart will help you to figure out what will grow in your area. For example, I know that I cannot very successfully grow watermelons in our area. They require too many frost free days to mature and they like warm nights in the summer and we have cool nights.

Once you have your list of what you want to grow you need to decide how often you want to eat that food. For example, our family likes to eat tomatoes 2 times a week just fresh sliced with dinner. There are 4 people in our family and each person eats about ½ a pound of tomatoes a meal. That would mean my family would like to have at least 4 pounds per week. In a high performance garden each tomato plant will produce about 2 pounds per week, once they get going. So if we want to have 4 pounds per week then we would want to plant 2 tomato plants.

I know that most of you are scratching your head trying to figure out how to guess the number of plants to grow and what your production will be like. It is common for low performance gardens yields to be inconsistent. So you cannot know what to grow because your yields are so varied. Boom one year and bust the next or just bust all the time. The consistency comes when you set up a growing system for your garden. In my greenhouses and gardens I have a chart where I can figure out on paper just what I need to plant to supply all my customers for the entire year. This is not difficult to do once you have a growing system set up in your garden. It is actually quite easy to figure out and is fun to do.

Below is the chart that shows the estimates of how many plants of each crop it would take to feed two people in our high performance garden system. Remember these are estimates for plants that are growing in ideal conditions and this may not equate to a low performance garden. If I were to plant a garden for 2 people I would put in this amount of these vegetables to give good sized servings each week that the plant is producing.

You don't have to put in all the vegetables on the list, just the ones you want to grow.

Abundance Garden Planting Estimates for 2 People

Plant	# of plants
Basil	4
Beans, Dry	49
Beans, green	100
Beets	85
Broccoli	9
Cabbage	6
Carrots	400
Cauliflower	9
Celery	25
Chard, Swiss	12
Corn, Sweet	25

Cucumber	4
Garlic	85
Kale	9
Kale (small leaves)	49
Kohlrabi	50
Lettuce, Leaf	100
Melons	4
Mustard	49
Onions (bulb)	85
Onions (bunching)	100
Peas	169
Peppers	6
Popcorn	25
Potato	25

Pumpkin	4
Radish	100
Spinach	50
Squash, Summer	4
Squash, Winter	4
Sunflower	3
Sweet Potato	9
Tomato	4
Turnips (globe)	42
Turnips (salad)	100

How Much Space

Now that you know what you want to plant we need to figure out how much space it will take. This is another enjoyable part of planning. It is all simple equations. We get out a map of our gardens and write on it what plants will go where. Once you have decided on your growing system, the system will have the spacing charts for each plant. Chapter 13 has the spacing chart for my high performance gardens. If you are going to garden in a low performance garden you can use the spacing on the back of the seed packages. Once you know what will go where you will know how big to make the garden. So now our planning has led us to the size of the garden. This is how high performance gardeners know how big to make their gardens. It saves a lot of time and energy to only grow the amount of plants that you want.

How Many Seeds and Plants

Good planning also saves you buying too many seeds and starter plants. With your garden plan you will know the exact number of seeds and starts that will go into your garden. For example: I know that for my family I will only need 12 tomato plants instead of 4 to satisfy my family's tomato cravings!

Many people will buy a package of lettuce seeds (somewhere in the 300 seed range). Go out to their garden and pour the seeds in a row until the package is empty. This is a waste of seeds and money. Your family will more than likely only eat one head of lettuce a week. The 300 seeds that were just planted will be too crowded and not all of them will grow. Let's say 100 live, the ones that do grow will mature within 3 weeks of each other so your family will have over 30 heads of lettuce for 3 weeks in a row and be without lettuce the rest of the season. This problem can easily be resolved with a plan. When you are shopping around for a high performance garden system for your garden be sure it has plenty of information on planning the garden. If it doesn't you will not get high production or steady all season production from your garden.

It would be nicer to have one or 2 lettuce plants all season long than to have 100 lettuce plants in a three week period. Lots of people claim that they cannot grow lettuce in the heat of the summer but it can be done. Our greenhouses have summer temperatures reaching 105 degree Fahrenheit every afternoon and we grow a steady amount of lettuce all year round. The overabundance of lettuce and the lack of lettuce the rest of the year is a planning problem not a growing problem. This is easily solved with a high performance education and a few new growing techniques.

If you like to preserve your food, the amount of extra food you will want to grow can be plugged into your plan and you will know what you will have to preserve. For my family the grocery store is optional because we do plan what we want to preserve for the winter. Through the off season we primarily eat from our pantry, root cellar and freezer. It is a great goal to have to primarily eat from your garden and you can do it if you take it one step at a time. Knowing what type of yields your garden is going to produce is the first step.

Keep a Journal

If you will keep a journal of what you planted, how those plants performed and what yield you got you will understand the patterns of your environment. With the knowledge you get from keeping a journal you will learn how to customize your garden for your yard. Now we are getting into the fun part of playing with your garden system. Once you are proficient at your garden, which can happen when you have a consistent system, then gardening takes on a whole new level of joy , excitement and fun.

The garden plan is essential for a great garden experience. Once you know what to grow, when to plant and how much you will get then the growing is easy.

CHAPTER ELEVEN
SUCCESSION PLANTING

If you have your garden already set up and there is no need to wait to rototill before you plant, you can begin as soon as the weather will allow. With a few frost protection systems you can start planting 4 weeks before your last frost date and comfortably grow 4 weeks after your first frost date in the fall. This means that warm climate gardeners can grow year round if they want to. Temperate climate gardeners can grow somewhere around 30-40 weeks of the year. The northern growers will be able to grow somewhere around 24-30 weeks out of the year. A high performance garden system can really extend the growing season. With the opportunity of a longer growing season we will need to learn how to grow our plants in succession.

Succession planting means we will plant crops in a specified time frame so we don't run out of harvestable food throughout the season. Leaf Lettuce is the best example. Most families like to have 1-2 pounds of lettuce to eat every week. They are not too interested in eating 30 pounds of lettuce in one week, yet that is how most people plant their lettuce.

Leaf lettuce will germinate in 1 week and be ready to harvest the outer leaves in 4 weeks from planting. The lettuce plant will produce lettuce leaves for harvest for about 10-12 weeks. Then the plant bolts and goes to seed at this time the leaves become bitter. Knowing these facts about leaf lettuce it is possible to create a plan to have some lettuce ready to eat every week. The high performance gardener knows the schedule of their plants production and works with this schedule to create food for the table in the right timing and volume.

In a high performance garden, the season is longer so it is important to grow multiple crops to keep the food rolling in. I learned how to do this by selling salad mixes to my local grocery store. The store does not want to hear that the lettuce has bolted and it will be 4 weeks before the next delivery. I have discovered how to plant the lettuce so we have lettuce to deliver all year round, every week without fail. I know it is possible for me to grow year round and to control the volume. This is easy to learn and your high performance system that you follow should have instructions on how to set up a secession planting schedule.

There are plants that take a long time to produce like tomatoes. Unless you live in the tropics your tomatoes will not go on a succession schedule. I have found with the longer growing seasons that I have created in my greenhouses that zucchini and cucumbers do better if they are on a one time mid-season replant succession.

Sometimes the plant will reach the end of its cycle and will not produce any more, like green beans. When this happens it is nice to know when to take the plant out and rotate in a new plant. A good succession plan will provide that information for you. This way your garden beds are always planted and creating more food. An idle bed is usually not given much attention and this is where the trouble begins. The weeds grow and things become ugly. If your beds are always planted you will be paying attention to them.

Succession planting is really easy once you know a few facts about your plants. This is a great skill to work on once you have your growing skills down. The first year just try the leaf lettuce succession. Once you understand leaf lettuce then add another plant and so on until each of your garden beds are always in use. When you can make your growing season longer by using raised beds, you here.

CHAPTER TWELVE
EXTENDING THE GROWING SEASON

When you can make your growing season longer by using raised beds, you save money on groceries and eat more nutrient dense food. There are several ways to extend your season. You can cover the plants and grow them longer in the garden. You can grow extra food and preserve it by putting it in the root cellar, freezing, canning or dehydrating your harvest.

Planting Early and Staying Late

In a raised bed system you don't have to wait for the garden to dry out to plant. So you can begin your garden 4 weeks before your last frost date in the spring. Once you know your planting zone and have figured out when your last average frost will occur you can count back 4 weeks and start your cool weather plants. The warm weather plants will have to wait until your frost date has passed.

Cool weather plants		Warm weather plants
Beets	Garlic	Basil
Bok Choy	Kale	Beans
Broccoli	Leeks	Corn
Brussels Sprouts	Lettuce	Cucumbers
	Mustards	Eggplant
Cabbage	Onions	Melons
Carrots	Peas	Okra
Cauliflower	Potatoes	Peppers
Celery	Radishes	Pumpkins
Cilantro	Spinach	Squash
Dill		Sweet Potato
		Tomatoes

Covering Protection Systems

If you are starting early you will need a protection system to save your plants from a harsh wind or a devastating frost. When you begin 4 weeks early and extend the season 4 weeks beyond your first frost date in the fall you will need some plant protection.

I have found that the last month of frost in the spring is usually not that hard of a frost. I watch the weather forecast and if a night is predicted to go below 28 degrees Fahrenheit then I cover the cool weather plants with a blanket and plastic on top of the blanket if it is going to rain or snow. The cover will need to be kept off of the plants with some type of support structure. You will also need some rocks or blocks to put on the edges of the blanket to keep it from blowing away. Go to our website http://thelivingfarm.org/project/tomatotrellis/ for videos on how to make an easy trellis with no tools, and how to protect your plants from frost. The easy trellis will double as a trellis system for your plants and as a type of cold frame to keep the frost and wind out.

For the warm weather plants, on my average last frost date I check the 10 day forecast for my area. If I see no frost predicted then I plant my warm weather crops. I continue to check the weather forecast and if it is predicted to go below 35 degrees Fahrenheit, I cover my warm weather plants. If it is going to rain or snow, I cover the blanket with plastic. Remember to use some blocks or rocks to keep the wind from taking your covers off.

Cold Frames and Heat Cables

If you are going to go to the extreme to extend your growing season you can grow in cold frames. These are small structures that you can put over your beds to create a small greenhouse to keep your plants alive. Heat cables or old fashioned Christmas lights (the kind that gets hot) can also be placed in your cold frames to keep the plants alive.

Plant Selection

If you pay attention to the descriptions of the seeds you are buying you can get plants that are more cold hardy then others. For example, I grow a lettuce called Winter Density and it will take more frost than any other lettuce that I grow. It does not do well in the heat and will give out on me if I try to grow it in my greenhouses at 105 degrees Fahrenheit in the summer. So I switch to another wonderful lettuce plant called Mottistone that thrives in the heat and produces beautifully. Because I grow lettuce 52 weeks out of the year, I switch out the Winter Density in the spring and grow Mottistone through the summer and fall and then I switch back to Winter Density for the winter and spring.

There are corn seeds that do better in cold soils and tomatoes that will set fruit when it is colder. So being aware of the plant varieties can really make a difference to your yields.

Winter Planning

Another way to extend your season is to leave your root crops in the garden over the winter. Beets, parsnips, onions and carrots can all take a frost and can be left in the garden to be harvested through the winter in zones 7 and warmer. You will want to put a layer of mulch over the crop to protect it from a hard freeze. In colder climates it is really hard to keep the frost from settling into the ground. I live in zone 4 and I take all the food out of the garden and hold it in the root caller or preserve it for future meals.

The other way that you can extend your season is to preserve some food for the winter. Planning to grow extra food that you can store in the root cellar, freeze, can or dehydrate for the off season is a great way to extend your growing season. There are preserving videos on our website a http://thelivingfarm.org/homegrown-preserving/ I will continue to add more training videos as the summer progresses. If you can boil water you can learn to preserve food, it is pretty easy!

With a little knowledge and ingenuity it will be easy for you to get another 8 weeks of food out of your garden and have food all year round from the pantry, root cellar or freezer.

CHAPTER THIRTEEN
UTILIZE ALL THE SPACE

High Performance Gardens utilize all the space available. If you are growing in a high performance garden then you will use tight spacing and vertical growing to save space and reduce weeds. I love my high performance garden because it can be so small and produce so much food. The planning system for the high performance garden is what keeps the garden small but powerful. There are several ways to save space and to leave little space for weeds. First there is tight planting, second is keep the beds planted all season including 4 weeks before and 4 weeks after your growing season. Finally, there is vertical growing.

Tight Spacing in Your Garden Beds

A super productive garden starts with tightly planting your plants. If you select plants that are smaller when mature then you can fit more in your garden bed. I grow a miniature cabbage called Stonehead in my gardens. It is a green cabbage that is planted 12 inches apart. In a high performance raised bed I can plant all the way across the bed and leave no room for weeds. Also I can stagger the rows to create even tighter spacing. The Stonehead cabbage is wonderful because one half of a head makes a coleslaw salad for my family. I can also make sauerkraut out of it if I want too. I have grown larger cabbages in the past but they take up a lot of bed space, the heads are too big for dinner and we tend to waste more of it. Tight spacing is perfect for raised beds that are not over 4 feet wide. When you have a bed that is under 4 feet wide and you have walking aisles all the way around it you can easily reach in 2 feet to take care of the plants and to harvest. I have one raised bed that is 5 feet wide and is just too wide to plant in the middle. I ended up putting a stepping stone in the middle of the bed so I could step in to work the plants.

The chart below shows the dimensions I use for spacing in my gardens. This will probably not work in a low performance row type garden. This plant spacing is intended for high performance raised beds with walking aisles all around. These plants are planted in blocks not long rows, there is no space in the beds between the plants. Everything is butted up to the next plant. All of the spacing of the plants is based on the plants final mature size not the baby plant size. The spacing is for how far apart to plant in all directions.

Plant	Spacing
Beets	3 inches
Broccoli	13 inches
Cabbage	13 inches
Carrots	2 inches

Cauliflower	13 inches
Celery	6 inches
Chard, Swiss	8 inches
Garlic	3 inches
Kale	13 inches
Kale (small leaves)	6 inches
Kohlrabi	4 inches
Lettuce	4 inches
Mustard	6 inches
Onions (bulb)	3 inches
Onions (bunching)	2 inches
Peas	3 inches
Potatoes	8 inches
Radish	2 inches
Spinach	4 inches
Turnips (globe)	3 inches
Turnips (salad)	2 inches
Beans, green and pole	4 inches
Corn	8 inches
Cucumber	13 inches
Melons	20 inches
Peppers	13 inches
Popcorn	8 inches
Pumpkin	1 hill (4 plants) 48"
Squash, Summer	1 hill (4 plants) 48"
Squash, Winter	1 hill (4 plants) 48"
Sunflower	13 inches
Sweet Potato	13 inches
Tomato	20 inches

Keeping the Beds Full

We covered this topic in Chapter 11: Succession Planting. It is important to keep the garden full of plants that are productive and grow well during the current weather conditions. I will start the garden with cool weather plants, get a crop then switch some of them out for warm weather plants. Another thing that I will do is to plant a fast growing crop around a slow growing crop. This way the fast growing crop can be making food while I wait for the slow growing crop to fill out and mature. A perfect example is how I grow a crop of spinach in the same section that I plant my zucchini. Place the zucchini in the middle and plant the spinach around the zucchini. When the zucchini gets so big that it wants all the space the weather is usually hot and the spinach is done.

Another thing you can do is to plan on a cool weather crop at the end of the season to take up the space that your warm weather crops occupy in the heat of the summer. Southern growers can have food all year round with this method. The garden will be full and productive for a longer time. If the garden is easy you will not mind having the garden active for a long time. It saves you money and is fun.

Vertical Growing

This is a fun topic and saves so much space in the garden. The advantages of trellising your plants so they grow vertically is that: you don't have to weed the parts of the plant that are in the air, you can grow more food per square foot of garden, your plants look beautiful, the trellised plants can shade your plants that don't like the full summer sun, and it is easier to pick the vegetables and the plants don't get as many diseases or pests.

Here are some plants that you can grow vertically on a trellis.

Cucumbers I grow all my cucumbers on trellises. I use my tomato clips to hold them to a string then clip the string to the trellis with another tomato clip. When the plants reach the top of the trellis, I just let it flow over the trellis and it will head back to the ground. I will trim off the suckers on its way up but I just let it go and leave the suckers as it is coming down. When I grow in the greenhouse and have a long season I will replant the cucumbers once to refresh the crop if necessary.

Melon It is possible to grow your cantaloupe and your smaller melons on your trellis. When melons are ripe they can detach themselves from the plant. I would advise putting the melon in a panty hose sling to hold up the fruit. I have never grown watermelons this way because we have trouble with them producing fruit because of our cold Colorado nights. I hear that it can be done but you will need to use a panty hose as a sling to support your crop.

Nasturtium

This an edible flower that also produces spicy edible leaves. This is a great plant to add color to your garden as well as a flower and a new leaf for your salads. Nasturtiums come in short bush variety and in long vines. I grow the long vines because I can grow them vertically and save on bed space.

Peas

There are two types of pea plants that you can grow. Some are short and need no trellising and the others can grow to over 6 feet long and will need to be trellised. The latter will produce more peas and is a great choice for production if you choose to trellis. When they are trellised you can grow a lot of food in a small space.

Pole Beans

Pole beans are like bush green beans except they grow very long vines. They will grow up a string, corn stock or on plastic netting.

Tomatoes

I love to grow my tomatoes vertically. It is so important to keep the fruit up off the soil so the bugs don't get them first. I grow mine as a single stem tomato plants and we will space the tomato plants about every 20 inches. Our typical production is between 30-40 pounds per plant per season.

Sweet Potatoes

I plant the slips 12 inches apart then trellis them up and over the trellis and let them hang pack to the ground. I get great potatoes and they don't take up all the space. Because I live at 5700 feet in the mountains I have to grow my sweet potatoes inside my greenhouse and space in there is a premium and vertical is the only way to grow them.

Winter Squash

Believe it or not you can grow butternut, acorns and other small winter squash on the trellis. Most will hang just fine by themselves but if you are nervous you can put them in a sling like the panty hose sling that is tied to the trellis for support.

A high performance garden will make use of all the space to increase your yield and reduce the weeds. These skills are easy to learn. If you want to learn about our trellising methods you can watch our training videos on trellising tomatoes online at http://thelivingfarm.org/project/tomato-trellis/

CHAPTER FOURTEEN
CREATING HUGE YIELDS

A high performance garden will yield $15-20 per square foot over the course of the season if done right. In my 128 square foot (4x32 ft.) Abundance Demonstration Garden and with just 15 minutes spent in it per day, I produced enough food for 2-3 people plus had some extra to give away and preserve for the winter. The yield was right at $20 per square foot.

Contrast this with the average low performance gardener. In 2009 the National Gardening Association published the *Impact of Home and Community Gardening in America*, in this study it was noted that a well taken care of garden would average ½ a pound of produce per square foot. That's just $2.00 in market value! These high performance garden systems truly take garden production to a new level.

Plants are amazing and given the chance they can grow really big and produce lots of food. Most people never get to see their plants produce amazing yields because they stunt their growth by growing the plants with low performance methods. When you grow your plants in poor soil that doesn't have the nutrients or the microbes to deliver the nutrients, the plants will not be able to reach their full genetic potential. Once you switch your soil to the rich soil that has the proper nutrients and happy microbial activity, hold on to your hat because you will see amazing results.

In the high performance garden we have stacked the odds in our favor so the plants can grow to their full genetic potential. I am still amazed after decades of gardening by what the plants do. Our 5 cucumber plants in our Abundance Demonstration Garden produced 216 pounds of cucumbers in 21 weeks of production. I made a lot of pickles and dehydrated cucumber slices so I could enjoy this harvest all year. The spinach made leaves that were bigger then my hand. Some of the spinach almost grew over the top of the zucchini plant! That is happy spinach and very high yields. I put up a lot of spinach in the freezer and enjoyed it all winter.

In a high performance garden we take special care to create the right type of soil, keep it organic, feed the microbes and create the perfect soil environment for our plants. We also take special care to protect the plants from the elements, bugs and diseases that can stunt their growth. By protecting our plants we will also increase our yields. Seed variety will also increase our yields as well as our watering techniques and weed control.

It is all about the dance. The interweaving of all 12 high performance characteristics will determine if your harvest is small or large. Each characteristic that is not achieved will affect the yield of your plants. The goal is to keep your eye on the game and keep the plants and microbes happy by following a system that achieves each characteristic of a high performance garden.

The more attention you give to the garden the better the production. If you have a garden that is easy on you and looks awesome you will spend more time in it. If you are observant while you are in your garden then your yields will be higher because you can stop an invasion or problem before it starts.

With a small daily time commitment it is possible to grow enough food for your family. Pretty incredible isn't it? If you encourage the other eaters in your family to join in on the gardening fun it will take you even less time. This is also fun to do with friends or neighbors. Sharing your high performance garden with others by having them garden alongside you is one of the priceless joys of high performance gardening.

CHAPTER FIFTEEN
FUN, REWARDING & ENJOYABLE

A high performance garden will be beautiful, fun, productive and so enjoyable. You will find the neighbors looking over the fence and asking you how you grew such an amazing garden. Your garden will be your pride and joy! It will also become your place that you go for peace and tranquility.

The garden is full of gifts beyond just the delicious food it produces. The decades I have spent gardening have been full of different garden gifts.

One of those gifts the garden has given me is pride. Even after 30 years I am still in awe of what the garden and I can co create. I get to be a part of the miracle of creation every day. I carefully harvest the most amazing vegetables then bring them home to my beloved family. I say to them, "Look at what I grew", with a smile as big as my vegetable, my chest exploding with pride. This is one of the best feelings in the world and I co create this with my garden.

Another gift my garden has given me is thoughtfulness. After asking the question of how to make the garden easier, more productive and take less time, I started thinking, researching, processing and planning. After the many challenges the garden presented me with, I was now ready to try a new style of gardening. This system showed promise of higher yields, easier chores, and time savings. Because of this thoughtfulness I have the high performance garden systems that I love today.

One of my favorite gifts that the garden has given me, and that I know it will give you, is peace. In a high performance garden system our gardens are weed free, just the right size, produces an abundance of nutrient dense organic vegetables, beautiful, a joy to work in , and sit in to observe. My garden gives me peace and tranquility as well as a delicious dinner.

There is magic that happens once you have become proficient in your high performance garden. As you become familiar with your plants and the system your garden truly becomes a high performance garden. Once you are proficient the whole garden world opens up to you. By having a system that you fully understand you now know what to expect, how to handle problems and your garden becomes fun and reliable. You count on your garden to feed your family year in and year out. This is where your food independence begins. Once you are proficient high performance gardener, you will have the ability to feed your family independent of the industrial food system.

Gardening is contagious. Once you get your system down the neighbors will start looking over the fence to take a peek at your lush garden. You will become the gardening expert of the neighborhood. Almost all of your neighbors will either be too scared to garden or are gardening in a low performance system. They will come to you for advice because they will want the garden that you have. I would like you to teach them how to grow food just like I have taught you. It is so important that we pass this knowledge along and keep people engaged in gardening and food independence.

What would our world look like if everyone grew most of their own nutrient dense foods right in their own backyard? What if these people shared their extra produce with those who couldn't grow their own? Just think of the impact we could have on the world's food supply and our environment if everyone that could, would grow a garden in their backyard.

This is my dream for the world, that everyone grows the majority of their own food in their backyard. When this shift begins, the first effect we would enjoy would be a healthier society. We would not be our burdening health care system and health care costs would go down. I see people living longer with more vitality than they ever could have imagined. People would experience less stress and more peace in their lives. I see older people spending time with younger people teaching them how to grow their own food. They would be taught to give a portion of the food they have grown to someone else who might not have the opportunity to grow their own. People would share their stories of how they grew their food, their chest swelling with pride as they teach others how to do the same. I see the teachers glow with satisfaction as their students learn to grow proficiently and then turn around and share their knowledge with the next generation.

This would also impact our planet and create a cleaner environment for all of us. With people growing the majority of their food in their backyards, there will be fewer trucks on the road, less pollution, less packaging, less food waste and less for the landfill. I see cleaner air with less pollution and less carbon dioxide as the plants carry $Co2$ down into their roots to store it in the earth.

Moms and dads with smaller food bills will not have to work as many hours to put food on the table. Instead they can be out in the garden with their children, teaching them how to grow amazing food. They would be taking family vacations with the money they saved by growing their own food.

I see more smiles as people enjoy the garden. People will spend more time outside away from the computer. They will be learning what it feels like to create something real; not something that is deleted with the press of a button.

Would you like to be a part of this new world? It all starts with you, in your back yard. Once you have learned to garden then share this with your friends and family and help others get their garden going in their backyard. This is bigger than just growing food; it is about helping others and our planet.

I am proud to be the catalyst that starts this new revolution. When you join us you will also feel the pride of doing something good for this world. Wow, how amazing would that feel? You can change the world starting in your own backyard then helping others to do the same in their backyards. This is how we can create a more sustainable future, one garden at a time. Go out today and get your garden started! First learn the new skills you will need to garden the high performance way. Then build your beds and when the time is just right, plant the garden. As you care for it, love it you will watch it thrive and create an abundance of joy, satisfaction and amazing nutrient dense organic food. My goal is to create satisfaction, health and vitality everywhere my teachings come in contact with another person. This is my legacy, a more beautiful world, one inspired gardener at a time.

CHAPTER SIXTEEN
YOUR LIFE'S NEXT CHAPTER

Now is the time to create the high performance garden of your dreams. It is time for you to become a part of the solution for a healthier planet and society. Now that you know all the components of the high performance garden, it is time for you to find a system that you like to grow in. There are several different high performance systems out there. I have recorded my system into a course called The Abundance Garden. When you go shopping for a system and a garden coach that you would like to learn from, make sure that the system has all 12 of the characteristics, and that you will get support throughout the growing season.

This is where the dance starts. As you combine all the different criteria of a high performance garden a system emerges that will delight and create amazing food. This is where gardening goes from a chore to a passion. Once you are proficient in the high performance skills needed your garden becomes so enjoyable. You will start honing in on your unique high performance garden.

Imagine a high performance garden in your backyard or front yard. You will be amazed at what you can grow. This will be a whole new chapter in your life. Not only will you be learning how to grow amazing nutrient dense food for your family, you will also be sharing what you learn with other families. You will be a part of the solution to help people and our planet to thrive again.

Now that you know the components of a high performance garden you will never be able to grow in a low performance garden without wondering why you put up with the struggle, it is not necessary. It is time to shed your old ways and start anew.

Here are the nine high performance practices you need to implement in your garden to begin the transformation to a high performance garden.

1. Make a plan: It is important for you to figure out what you are going to grow before you buy a single plant or seed. If you make the garden too big it will be too much work and not enough fun. Grow only what you are going to eat.

2. Vow to always grow organic: Now that you are aware of the microbes growing in your garden and how that affects your life and the life of the planet be respectful and vow to take care of all life by always growing organic.

3. Add minerals: Keep your soil mineralized so the microbes and plants can keep you mineralized.

4. Don't walk on your soil: protect the fragile microbes.

5. Make compost and compost tea: Learn how to recycle your kitchen and yard waste to create food.

6. Mulch your garden and plant tightly to reduce the weeds.

7. Use succession planting to keep the garden full.

8. Learn how to use plant protection to extend your season.

9. Enjoy your garden and share your crops and knowledge with another person.

If you would like support in any one of these areas you can join the High Performance Garden Community. Becoming a part of the community is completely free and you will receive weekly High Performance Gardening Training Videos. Each training video focuses on a High Performance Garden Characteristic and specific skills that you will need to transform your garden into a high performance garden. You will also be inspired to grow the nutrient dense vegetables your body needs, share your harvest with those who cannot garden and teach those who can garden the high performance gardening methods. Come join today at http://thelivingfarm.org/high-performance-garden-community

Another free support in your gardening journey is the High Performance Garden Show. This show follows the 2016 growing season in my Abundance Garden demonstration garden. In this "real time" gardening show, follow a 128 square foot demonstration garden through a 33 week growing season. Learn easy weed-free, productive and nutrient dense gardening techniques. Garden alongside me as I try to grow $2990 worth of produce in 128 square feet in just 33 weeks. You will also have access to the bonus guide that will show you what to plant each week of the gardening show! You can start a garden at any time with the guides help. Sign up for free and watch the 2016 season at http://thelivingfarm.org/high-performance-garden-show/.

If you want to grow a high performance garden in one season you can enroll in the Abundance Garden Course. This online course is the most comprehensive high performance system online and available to beginners and professional gardeners alike. The beautiful thing about this course is that we will walk you through starting your garden and growing nutrient dense produce in your first season. Our training is so complete we can even tell you what day to start your garden and how long you can grow it for in your region (that is if you are in the US). You can find out more about the course and how to upgrade your garden in one season with our how to get started video go to http://thelivingfarm.org/abundance-garden-course

Got a garden question or need some garden advice? Send us an email to info@thelivingfarm.org to get gardening answers.

ABOUT THE AUTHOR

Lynn Gillespie is a 3rd generation farmer who has been developing her high performance garden style for decades. Her home is The Living Farm where she has raised the fourth generation of organic farmers to take on the farm and family businesses. This 210 acre family farm near the base of Mt. Lamborn in Colorado sustains The Living Farm Café, her active CSA and the local community. Tom Gillespie, her husband, was born on this land and has farmed since the time he was old enough to walk. Over the years she has run their greenhouses and grown bedding plants, trees, shrubs, fruits and vegetables. Her dairy sheep flock is one of her pride and joys. The flock provides milk for cheese making, and wool for spinning, weaving and felting. Her first gardening book was written in 1998, "How to Grow all the Vegetables you can eat Right in Your Own Backyard". Her second book, "Cinderblock Gardens", was written in 1999. In 2007 Lynn's dream of expanding The Living Farm's educational outreach was first realized. The Living Farm Sustainable Education Center was developed and teaches students from around the world how to become small scale organic farmers.

By 2009 The Living Farm released their first documentary, "Locavore; Local Diet Healthy Planet" with the goal of educating people on how eating locally can significantly affect your health, the health of the planet and stimulate your local economy. Lynn's life mission is to empower gardeners of all levels to high performance garden in an easy, fun, productive and always organic way. She developed the groundbreaking Abundance Garden Course to fulfill that life mission. The Abundance Garden Course guides students to build a high performance garden in one season. This course was launched at the end of 2014 and it is guiding gardeners from all over the nation to create their high performance garden. *To learn more go to www.thelivingfarm.org*

www.ingramcontent.com/pod-product-compliance
Lightning Source LLC
LaVergne TN
LVHW020936090426
835512LV00020B/3383